Mysteries of the QUANTUM UNIVERSE

by Thibault Damour
& Mathieu Burniat

Translated by Sarah-Louise Raillard

PENGUIN BOOKS

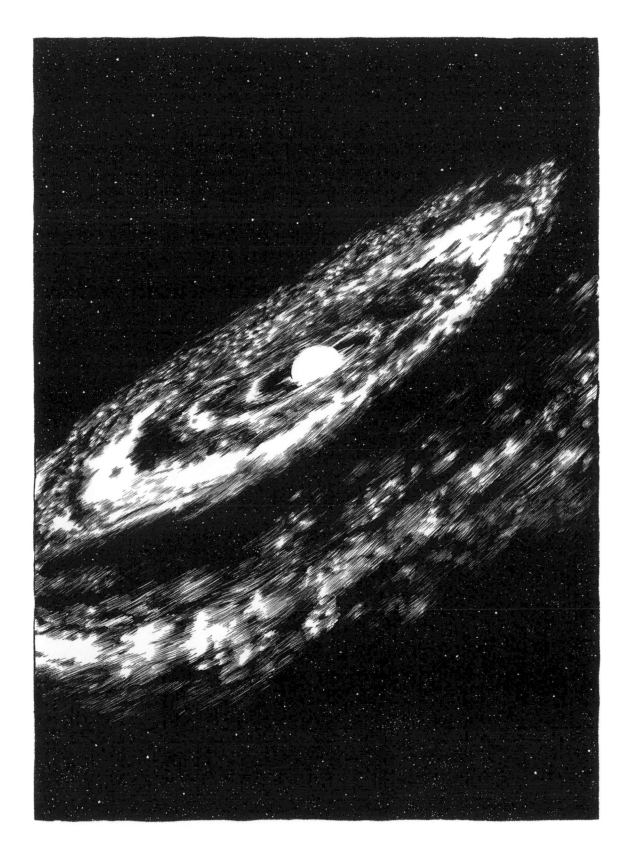

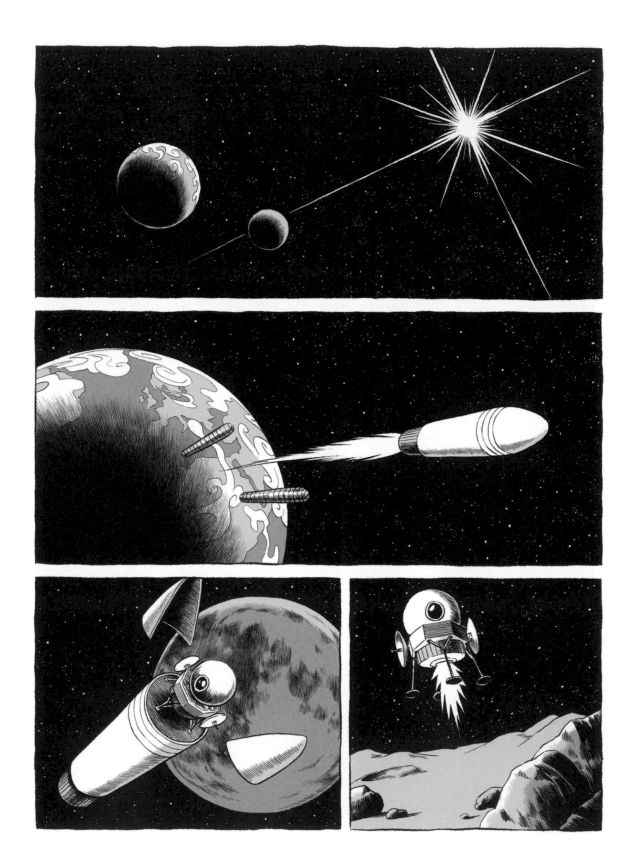

After 'Radiocarbon in the Holy Sepulchre'...

'Lobby Arcanum'...

And 'Neon Investigation'...

Our two heroes are investigating an unsolved mystery... on the moon!

The whole world waits with bated breath, hoping that these two wise minds will be able to...

I have to interrupt you, Jean-Claude!

It seems that Bob is about to exit the Paloma module.

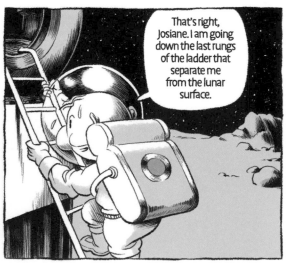

That's right, Josiane. I am going down the last rungs of the ladder that separate me from the lunar surface.

6

Mysteries of the quantum universe
In search of lost reality

Quantum theory describes the atomic world. A world made up of protons, electrons, photons and a whole slew of other particles. More than a century after its discovery, this theory remains mysterious, as it appears to bestow a number of surprising properties on the microscopic world, such as spatial delocalization and the superposition of conflicting realities! Quantum theory challenges our normal understanding of the macroscopic world that surrounds us. To discuss recent breakthroughs in quantum physics, a number of eminent international scientists will meet at the Hotel Metropole in Brussels, during the 25th Solvay Conference.

Opposite, the participants of the historical 5th Solvay Conference in 1927.

15

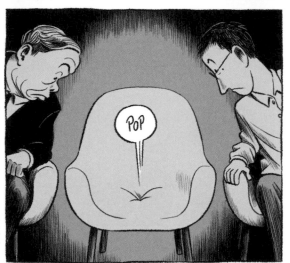

So what brings you here, mein Herr?

I was at a conference on quantum physics and well, I kind of got lost.

Not surprising! Most physicists can't find their way!

Uh-huh. Well, aren't I in a pickle...

That being said, you're lucky you crossed paths with me.

I'm Max Planck.

Uhm... I'm Bob.

I'm the one who opened the door to the quantum world without really knowing it.

Really?!

Uh-huh... by discovering, in 1900, a new fundamental physical constant...

A 'constant'??

Yes... a quantity that appears in the laws of nature...

19

Let's hope that h is nice enough to stay the same size!

Rest assured, it's a constant.

Its value is 0.000 000 000 000 000 000 000 000 0066. cm2 g/s.

That's quite small!

Size is entirely relative in physics.

How did you discover this constant?

At the time, I was trying to explain why fire is red, and why the Sun is yellow.

It's because of their temperature, right?

Exactly. These embers are at 800°C, more or less, and thus radiate light that has a frequency that predominantly corresponds to the colour red...

While the Sun, because of its higher temperature (5,500°C) looks yellow to us.

Lower frequencies ← → Higher frequencies

And you were the one who discovered all of this?

No! Some theoretical arguments and certain experimental results had already helped to approximately describe how thermal energy was distributed across different frequencies.

But we can't be satisfied with approximate and incomplete results...

I needed a full theoretical explanation of this law of distribution.

But this sort of calculation must be incredibly complicated!

Because in fact, you'd need to know in detail the constitution of each log, the distance separating them...

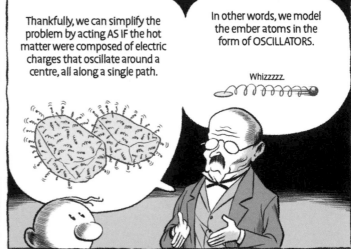

Thankfully, we can simplify the problem by acting AS IF the hot matter were composed of electric charges that oscillate around a centre, all along a single path.

In other words, we model the ember atoms in the form of OSCILLATORS.

Whizzzzz.

To represent an oscillator, imagine a small marble attached to a spring.

If I pull on the marble like so, I give it a certain amount of energy.

When I let go, the marble begins to oscillate to a greater or lesser degree, depending on the energy it received.

Boiiiing.

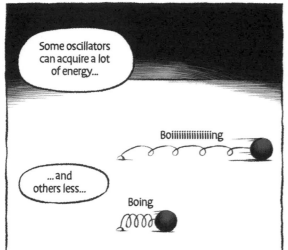

Some oscillators can acquire a lot of energy...

Boiiiiiiiiiiiiiiing

...and others less...

Boing

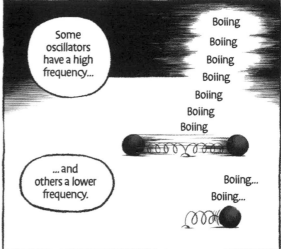

Some oscillators have a high frequency...

Boiing
Boiing
Boiing
Boiing
Boiing
Boiing
Boiing

...and others a lower frequency.

Boiing...
Boiing...

These oscillators are really useful to describe the characteristics of heated matter that emits light.

Hmmm... Energy...frequency....

Despite this simplification, though, there was still a problem...

The total energy of these embers is the sum of the energy of all of those oscillators....

Given that each oscillator can have a greater or lesser amount of energy....

Boiiiiiiing

Boing

...I had to count the number of ways in which the total energy of the matter could be distributed amongst all the oscillators!

So... Let's imagine a table of 10 rambunctious toddlers. They're our oscillators.

They are about to feast on Uncle Planck's delicious crepes...

The total energy of the matter is this bag of icing sugar.

I have to distribute it amongst all 10 children.

Well, the difficulty here is that I have to COUNT THE NUMBER OF DIFFERENT WAYS THAT THE SUGAR CAN BE DISTRIBUTED....

Option 1

51.04 grams for you... | 2.78 grams here... | 160.90 grams there... | 0.04 grams... | 75.33 grams... | 300.02 grams... | 12.72 grams... | 40.48 grams... | 89.75 grams... | 1.46 grams.

Option 2

0.034 | 437.08 | 15.53 | 4.102 | 39.13 | 138.84 | 3.657 | 70.97 | 1.508 | 15.26

Etc., etc.

22.77 | 0.837 | 160.92 | 37.112 | 0.0091 | 52.48 | 16.286 | 3.07 | 6.095 | 70.00...

But that's an infinite number! You can't count them all!

Boing Boing

Boing Boing

Boing

Boing Boing

Exactly. I was pulling my hair out. Finally, out of despair I opted for a rather unorthodox method...

And there you go!

SUGAR

Brilliant! More sugar!

...in cubes!

SUGAR

See, I acted AS IF energy only existed in the form of INDIVISIBLE pieces.

Suddenly, it's possible to count them all! For example, here I have 100 sugar cubes.

100 cubes for 10 children, that's....

4,263,421,511,271 different ways of distributing the sugar!

But... this method seems completely arbitrary!

How did you even decide on the size of your 'packets' of energy?

I'm getting to that! That's when I discovered 'h'...

In fact, I found that the size of my packets of energy...

... had to be proportional to the frequency of the oscillator in question.

Boing... Boing...

Boing Boing Boing Boing Boing Boing Boing Boing Boing

This result introduced a new constant of nature:

h is simply the relationship between

the size of the packets of energy...

and the frequency of the oscillator.

$$h = \frac{E}{f}$$

Hence my famous equation, E = hf !

Wow! That's impressive!

So if I've got this right...

... for your work, you modelled hot matter in the form of oscillators

and in order to be able to make certain calculations, you acted as if these oscillators could only have certain levels of energy....

They oscillate like this...

1hf

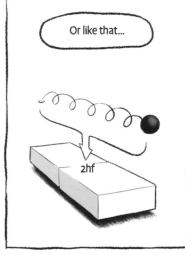

Or like that...

2hf

But not like this....

1,472 hf

Exactly.

Even if I don't really believe that an atom is like an oscillator that can only possess certain levels of energy —

my calculations nonetheless allowed me to obtain results that corresponded to what we can observe!

All thanks to the introduction of h!

Oh? Well?

Follow it, Bob. It will lead you to even greater discoveries.

You're only at the gates of the quantum universe.

Thanks for the crepe!

Ah, that's great.

No doubt you're familiar with Planck's calculation method, which segments the energy of matter?

Right... the sugar cubes... pretty clever, huh?

You said it! It's genius.

So ?

Planck said that he used this method for counting purposes, without believing that it was an accurate description of reality.

But what he finds difficult to admit is that the energy of matter is ACTUALLY segmented into little packets.

Nothing fictitious about that!

You mean to say that Planck's oscillators can ONLY exist in the energy states 1hf, 2hf, 3hf...

1hf   2hf   3hf

and not, for example, in energy state 2.374 hf?

2,374 hf

Exactly. And my discovery was not just limited to the theoretical oscillators used by Planck to model radiant matter...

I'm talking about EVERY MATERIAL SYSTEM capable of oscillating or vibrating!

How so?!?

A swing is another example of an oscillator.

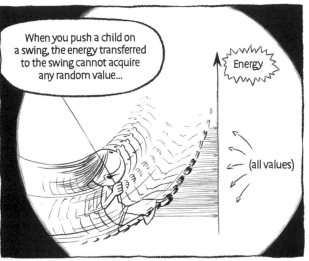

When you push a child on a swing, the energy transferred to the swing cannot acquire any random value...

Energy

(all values)

but only very specific levels!

Energy

In other words, energy is not CONTINUOUS, but DISCONTINUOUS.

You mean that when you swing, you can't reach every different height?

Precisely. But since we're talking about a microscopic oscillator, it's impossible to see this with the naked eye.

Do you understand?

Regardless of whether they are microscopic or macroscopic, oscillators can only exist in QUANTIZED energy states!

Now I understand the root of the word 'quantum'....

And you think that Planck had other preconceived notions?

Oh, yes...

Like most physicists, Planck was convinced that light could only be a CONTINUOUS WAVE propagating through space.

A number of experiments did indeed suggest this...

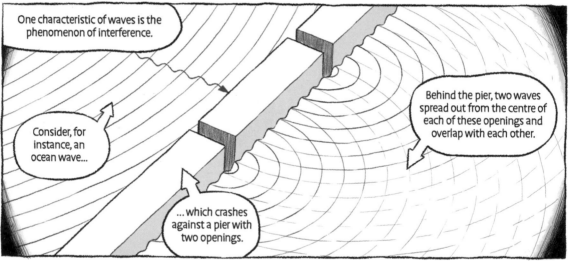

One characteristic of waves is the phenomenon of interference.

Consider, for instance, an ocean wave...

...which crashes against a pier with two openings.

Behind the pier, two waves spread out from the centre of each of these openings and overlap with each other.

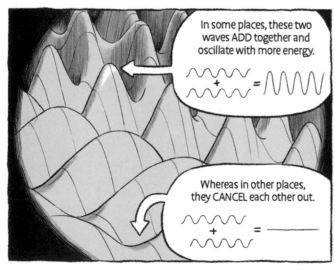

In some places, these two waves ADD together and oscillate with more energy.

Whereas in other places, they CANCEL each other out.

Light exhibits such interference effects.

Click

That's not all! I also showed that each particle of light has a fixed energy.

ε

This energy, dear Bob, can be obtained by multiplying the frequency of light by the constant h!

$$\varepsilon = hf$$

Oookay, wait a minute...

E =hf... that's exactly the size of the energy packets used by Planck!

And that I used to describe the energy of material oscillators!

Except that here, the equation means something different.

It states that LIGHT is itself composed of finite energy packets!

Oscillators, light...

Are you trying to quantize everything?!

Let's not stop there...

Let's turn to what happens inside MATTER...

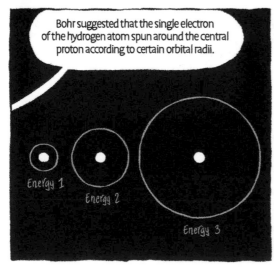

40

That means that an atom is like a solar system, where electrons orbit around a central nucleus?

Except that such a description can't be right...

Such a model would be completely unstable!

If an electron truly followed a circular orbit, it would emit light, like any other accelerated charged particle.

And it would rapidly lose energy.

Imagine that the electron is a low-altitude satellite...

By orbiting around the Earth, it would experience atmospheric friction. It would fall to lower and lower orbits, ultimately crashing into the ground.

So what does an atom look like, then?

We still don't know.

But what is certain is that Bohr's equation for the orbital radius of the first level of energy reveals something important about the size of atoms!

E_1

Click.

And since the structure of the world around us is based on the particular size of each atom, we can say that the APPEARANCE OF OUR WORLD RELIES ON THE EXISTENCE OF THE CONSTANT h !

We're so insignificant after all!

Bohr didn't stop with this description...

He also hypothesized that atoms sometimes jump ABRUPTLY from one energy level to another.

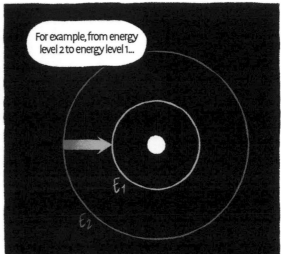

For example, from energy level 2 to energy level 1...

E_1

E_2

What makes an atom jump from one level to another?

We don't know!

So you're saying...

We don't know what an atom looks like, or why or when it changes energy states?!

Correct. And that's why I came up with an abstract method to describe atomic behaviour.

Since we don't know how or when the transition happens,

I suggested describing all of the possible atomic transitions in a PROBABILISTIC manner.

How so?

$[A_{nm}]=$

By creating a table where I wrote down all the different probabilities per second that an atom might move from one energy level to another.

??... And how many levels are there in an atom?

An infinite number!

E_1 E_2 $E_3...$

But then... the possibilities of an electron moving from one level to another are infinite!

And that's why my table is also infinite!

So... if this table represents probabilities... that means that you've incorporated chance into the description of an atom?

Exactly! In the absence of any explanation for transitions between atomic states, I have PROVISIONALLY told God to play dice!

Not only is your table abstract, but on top of that it only describes random events!

Yes, but it was nonetheless the first description of transitions between atomic states!

And as bizarre as it may seem to you, this description allowed me to learn more about quanta of light.

What's the relationship between atoms of MATTER and quanta of LIGHT?

I already knew that when an atom was bathed in thermal light radiation, it moved around in a random manner...

I then realized that atoms behaved like this because during an atomic transition, they had to randomly EXCHANGE MOMENTUM with light.

Atom in light

'Exchange momentum'?

Consider that our atom is a shooter caught in an ambush...

The quanta of light are the bullets that he shoots and receives.

When he shoots, he recoils from the MOMENTUM of the bullet shooting out.

Likewise, when our shooter is hit, he absorbs the MOMENTUM of that bullet.

Then imagine that he is constantly firing and being hit at the same time. From afar, an observer would see him moving around randomly.

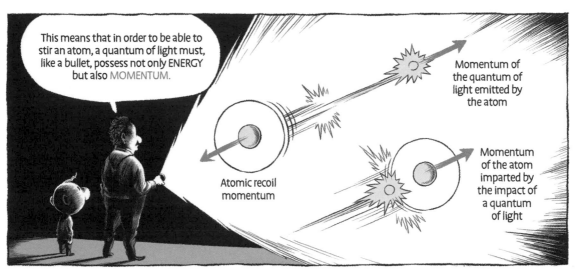

This means that in order to be able to stir an atom, a quantum of light must, like a bullet, possess not only ENERGY but also MOMENTUM.

Momentum of the quantum of light emitted by the atom

Atomic recoil momentum

Momentum of the atom imparted by the impact of a quantum of light

Energy... momentum...

If I understand correctly, a quantum of light is therefore quite similar to a moving particle of matter!

Exactly! But this idea makes the true nature of light even more mysterious.

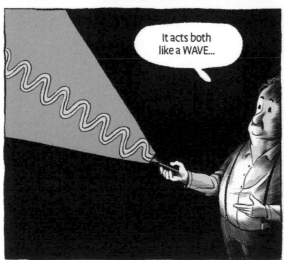

It acts both like a WAVE...

...and a stream of particles!

How can we reconcile these two completely different ideas?

Is there a theory that allows us to MERGE the two?

And we may wonder whether MATTER is really made of particles, as it is generally believed.

Oooh!

If you want to know more about this, it's best that I leave you in the care of a French physicist

who has just proposed an elegant synthesis between matter and light.

VROOOOM

Bob, may I introduce you to the Prince de Broglie.

Do come in, my dear sir, I insist.

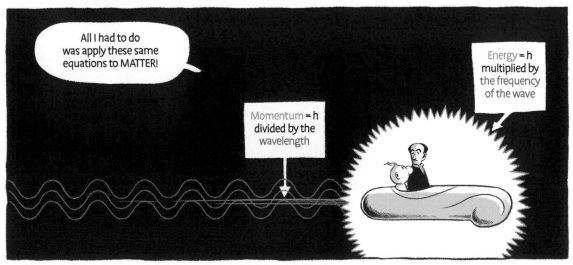

Hmm... Uh...

Does this concept of waves of matter help us understand atomic structure any better?

I think so. At least in the case of the hydrogen atom, which was what Mr Bohr focused on in his research.

Oh yeah? But...

I was told that it was incorrect to think of an atom as a solar system...

so I'm happy to have another way of imagining atoms!

In that case, please follow me.

See, the main idea is that the electron orbiting around the atomic nucleus is 'guided' by a wave of matter!

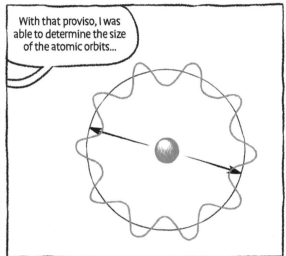

Thanks, Prince de Broglie, for taking the time to explain your work to me.

It was my pleasure.

Who would you suggest that I talk to now to continue my investigation?

Hmm, well... I heard about this 23-year-old German prodigy....

He's been lucky enough to work with great physicists like Bohr, Sommerfeld and Born...

His name is Werner Heisenberg.

Where could I find him?

I'm afraid I can't help you there.

Good luck, dear friend!

Slam

Shoot! What to do now?

What's Heisenberg doing on this godforsaken island?

The poor fellow had a serious bout of hay fever. He went to Heligoland to enjoy the sea breeze — and the lack of flowers.

It's thanks to this little trip that Heisenberg made the greatest breakthrough in quantum physics!

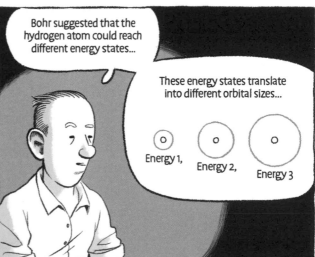

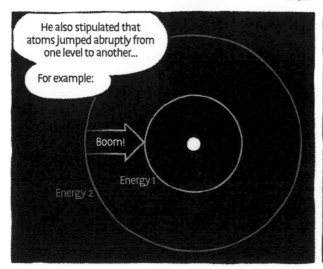

65

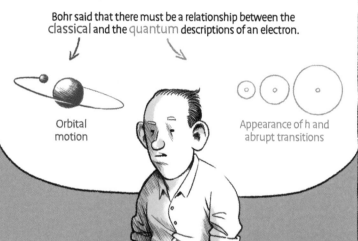

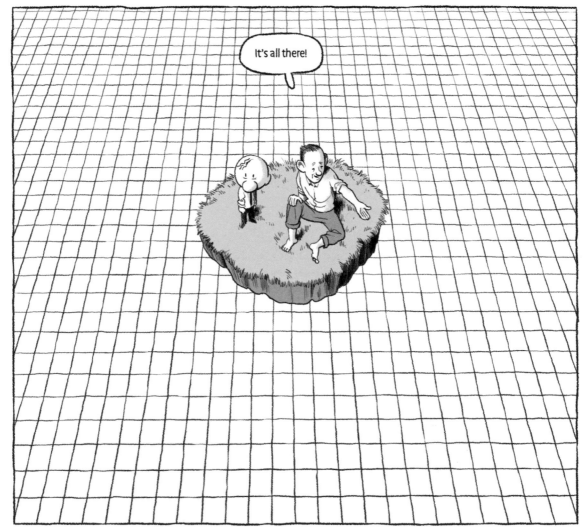

Aargh!!

Yes...

And that's when I lost my dog Rick.

In fact, I thought he was dead and stuffed. But he came back, I don't know how...

...thanks to the 'power of the quantum universe' he said.

Listen, there's not much I can do for your dog.

But in terms of your quest, I can certainly help you out.

I caught wind of Heisenberg's discovery. I'm a physicist myself.

Erwin Schrödinger.

Bless you.

If it's any reassurance, the transcendental nature of Heisenberg's matrix mechanics scares me too.

Oh yeah?

Even if the mathematical results are correct, Heisenberg refuses to think about what is hiding behind them.

According to his theory, the world of the infinitely small is impenetrable and abstract.

I decided to take another theoretical path.

You found another, more intuitive way to describe atomic physics?

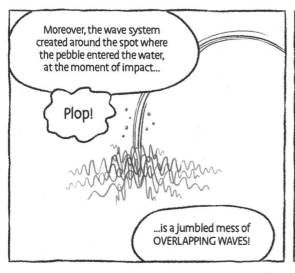

Moreover, the wave system created around the spot where the pebble entered the water, at the moment of impact...

Plop!

...is a jumbled mess of OVERLAPPING WAVES!

Yeah? But these same waves moving towards us seem so smooth and orderly...

That's what I wanted to get to.

The wave system naturally DECOMPOSES as it propagates into a sequence of waves which, at least locally, appear to have a simple structure composed of orderly crests.

This is because the jumble at the point of impact...

was made of the SUPERPOSITION of an infinity of simpler, plane waves.

So how does this observation help you to develop de Broglie's ideas?

He brilliantly established the mathematical formula relating the speed of a plane matter wave to its wavelength...

This, dear friend, is the notation I chose for the wave associated with any material system.*

Do you mean any particle of matter, like an electron inside of an atom?

That's what I thought when I first started my work...

In fact, I quickly realized that I had to associate a wave ψ not with each particle taken separately, but with ALL THE PARTICLES that make up a given system.

Ok, now I'm really lost.

Here's an example: let's take the second simplest element after hydrogen, helium.

The helium atom contains two electrons that 'orbit' around its nucleus.

For the sake of simplicity, let's imagine that the nucleus can be modelled as a single particle...

The helium atom then becomes a three-particle system.

* The Greek letter ψ (psi) is pronounced 'sigh'.

79

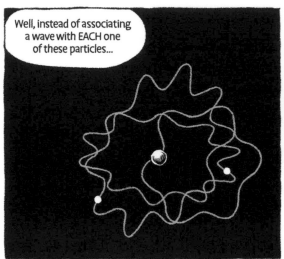

Well, instead of associating a wave with EACH one of these particles...

you have to associate a single wave with the whole SET of particles.

But... how do you find this wave?

My wave, which I called ψ, depends not only on time, but also on the variables representing the position of the three particles in space at the same time!

Gosh! That's one complicated wave, then.

And this is only for a very simple atom!

As strange as this wave ψ may seem, it helped me to come up with an equation that describes all of quantum physics!!

Look...

It was the first time this sequence of energy states was derived from a physically coherent theory!

So, does your solution include Bohr's circular orbits?

Not at all.

Do you at least have de Broglie's waves, spinning around the nucleus and guiding the electron?

Nope.

So what image of the atom does your equation give us?

In my opinion, we must completely abandon the idea that matter is composed of particles!

But then... what is an electron??

82

I suggest that the electron is a wave, and only a wave. And there is not, as de Broglie argued, both a wave and a particle.

And this wave, WHAT DOES IT LOOK LIKE???

If we look for example at level 1, the lowest energy level,

then the electron-wave doesn't resemble a wave spinning around a nucleus at a finite distance AT ALL...

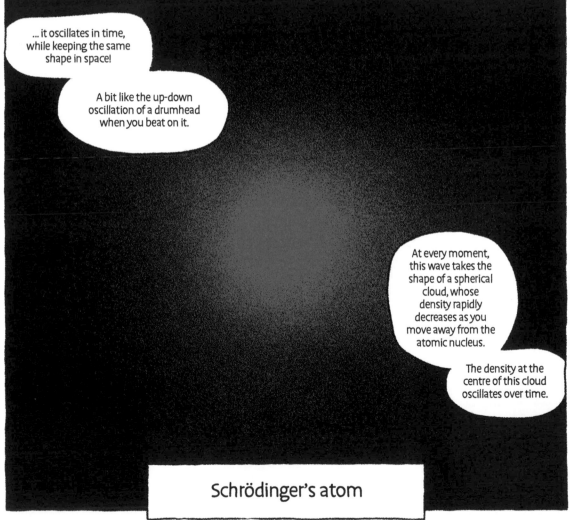

... it oscillates in time, while keeping the same shape in space!

A bit like the up-down oscillation of a drumhead when you beat on it.

At every moment, this wave takes the shape of a spherical cloud, whose density rapidly decreases as you move away from the atomic nucleus.

The density at the centre of this cloud oscillates over time.

Schrödinger's atom

For example, I should generalize my idea (which only applies to matter) into a formula that is capable of describing the quantum aspects of LIGHT.

Well, I know that I'm contradicting a lot of my contemporaries...

Oh! About that...

I am just about to meet some of them to discuss what quantum physics tells us about reality...

It should be quite interesting! Would you like to come along?

With pleasure!

Great, let's go!

?

What, you don't know the famous Max Born?

Let's not exaggerate....

Yes, yes.... a star is Born!

He he.

Ah, good one!

Ha! Luckily for us, this star doesn't agree with your farfetched ideas!

Max Born doesn't agree with my idea that the quantum world has a purely WAVE-LIKE nature.

Oooh.

That's right, Schrödinger believes that we can just scrap everything that is DISCONTINUOUS in quantum phenomena.

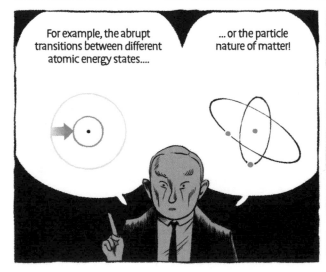

For example, the abrupt transitions between different atomic energy states....

... or the particle nature of matter!

So you reject Schrödinger's equation?

Not at all!

But I don't agree with how Schrödinger interpreted it to explain our reality.

For example, I looked at what Schrödinger's interpretation implied for the COLLISION between an electron and an atomic nucleus...

This ball is our electron.

And that one, our nucleus.

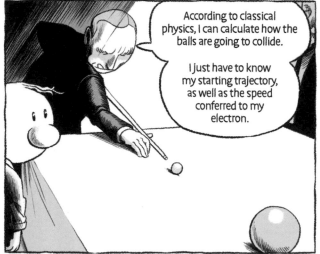

According to classical physics, I can calculate how the balls are going to collide.

I just have to know my starting trajectory, as well as the speed conferred to my electron.

Boom! Just like I predicted.

Schöner Schuss!

But I learned that the electron was not a classical object like a billiard ball but rather a... a thingy whose very nature remains shrouded in mystery.

Exactly right! Let's turn to Schrödinger's idea, according to which the particle-like electron is replaced by a wave-like electron...

93

This is where I think Schrödinger's theory is flawed...

He only looked at STANDING waves, describing electrons spread out inside of an ATOM.

But he didn't take into account the case of a MOVING electron, where its particle nature appears.

BEEP

Hold on, so you don't reject Schrödinger's equation...

But you don't agree with how he interpreted its results?

That is, that matter is entirely composed of waves?

That's right.

But then, what does this equation tell us about reality?

Since it brings into play a wave.

To answer this question, I'd like to start with an observation of Einstein's...

when he was working on light quanta, he told me several times that he thought that waves only served to GUIDE the quanta.

It's true... I spoke of a 'ghost field' that determined the probability of a quantum taking one path or another.

Well, in the case of matter as well, I believe that Schrödinger's wave ψ is a ghost field...

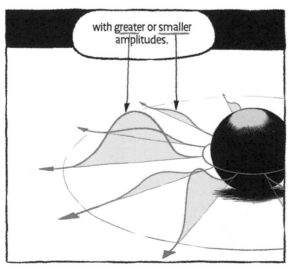

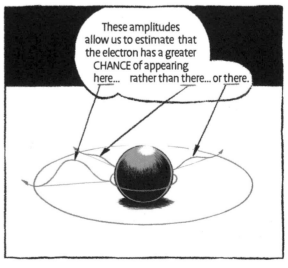

96

And when it hits the detectors...

BEEP

What? The amplitude of the wave ψ was greater at that point!

I thought the electron was going to show up there!

There was in fact a greater chance that the electron would show up there.

But there was still some probability that it would hit the detector in question... which is what happened.

25% CHANCE

2% CHANCE

5% CHANCE

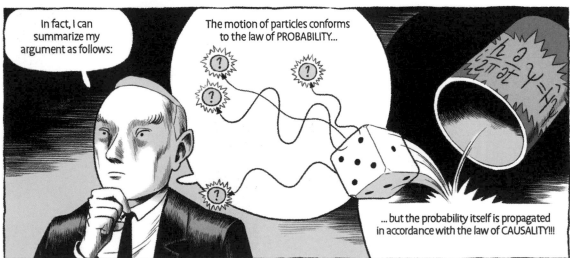

In fact, I can summarize my argument as follows:

The motion of particles conforms to the law of PROBABILITY...

?

?

?

?

... but the probability itself is propagated in accordance with the law of CAUSALITY!!!

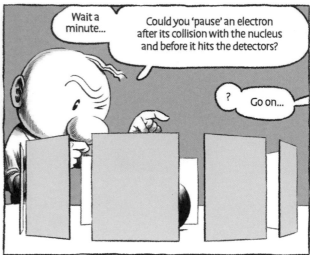

99

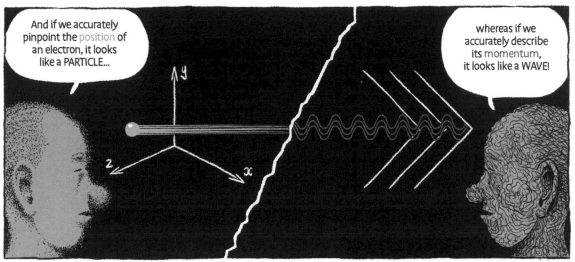

And if we accurately pinpoint the *position* of an electron, it looks like a PARTICLE...

whereas if we accurately describe its *momentum*, it looks like a WAVE!

In other words, the more you try to observe the particle nature of an electron, the less you'll be able to observe its wave-like nature.

And vice-versa.

Sniff

Bah! That doesn't really help me understand the intrinsic reality of atoms.

I already told you: there is no objective quantum reality!

We must give up trying to describe the atomic world in terms of space, time and causality!

We have to be content with reading the results recorded by our measuring apparata.

103

109

I'm Hugh Everett, a graduate student at Princeton...

For the past few months I've been trying to decipher what quantum physics tells us about the real world...

...and I think I've JUST found the answer!

You hear that, Bob?

We're all ears!

Are you familiar with Schrödinger's cat?

Uhm...

Like Einstein, Schrödinger believed that we couldn't separate quantum physics from the macroscopic world.

Also like Einstein, he was aware that his wave ψ exhibited flaws in our description of reality.

Schrödinger came up with a thought experiment involving a cat to illustrate one of these flaws.

To explain my idea, I would have to conduct this famous experiment.

Problem is, I don't have a cat on hand.

I volunteer!

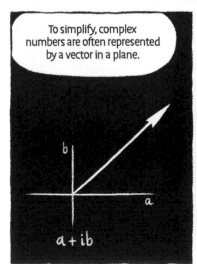

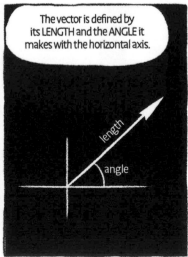

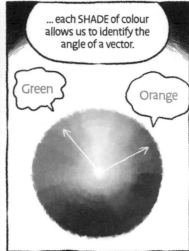

On the spatial coordinates of ALL the elements of a system!

You mean, the billions and billions of particles that make up the inside of this box?

Exactly!

We must therefore associate a complex number ψ with each spatial configuration.

Let's take the initial configuration for example: Rick just got into the box.

Like I explained, we can represent a complex number by a colour...

Shade of green

High intensity

And what is the value of ψ later on?

We get it by solving Schrödinger's equation, after including in the mathematical object Ĥ everything that contributes to the energy of our system.

$$i\frac{\hbar}{2\pi}\frac{\partial}{\partial t}\psi = \hat{H}\psi$$

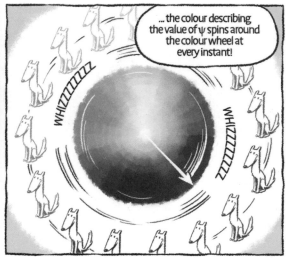

The outcome doesn't seem to be much different! There are still two Ricks!

... except that now, they're two different colours.

Exactly. Reality is a little bit like a movie made by superimposing several images over each other.

It seems like the blue Rick is brighter...?

The intensity of each situation depends on its likelihood of existing.

So if the alive Rick is brighter, it's because the radioactive sample I put in the box had a 2 out of 3 chance of not decaying after 5 minutes.

WAAAAH

In other words, alive Rick exists more than dead Rick.

Isn't that uplifting?

It's a little bit like all the waves emitted by different radio stations...

Depending on how you set up your receiver, you can hear a particular station, without being aware that the waves from the other stations are also present in space, overlaid right on top of yours.

What would happen if we were to turn the Geiger counter on again?

The living version of Rick would again split into two.

But... in the quantum world you're talking

... I imagine it's not just dogs and gases that are thus multiplied?

You're right, Bob. In fact, we're all part of this box...

WOOOOOOOOOOOOOOOOOOOZ

ZZWOOO

THE QUANTUM UNIVERSE IS EVERYWHERE!

OOOOOOOO

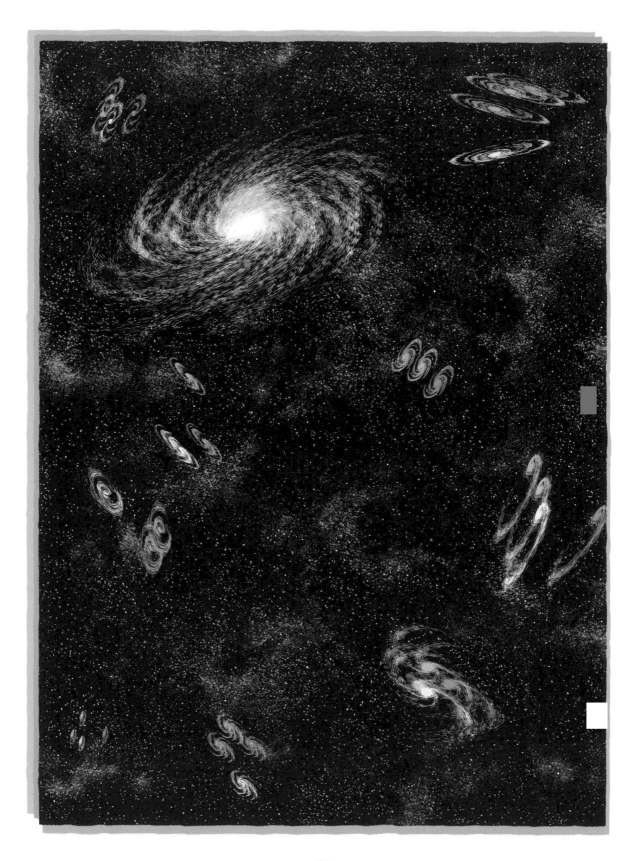

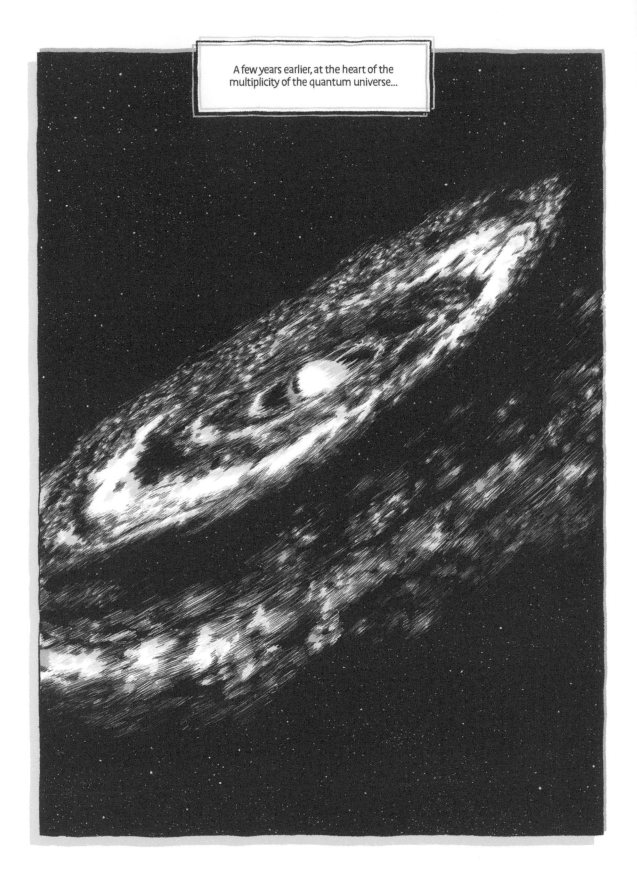

A few years earlier, at the heart of the multiplicity of the quantum universe...

GLOSSARY

These notes offer more details about the scientific ideas introduced in *Mysteries of the Quantum Universe.* It is certainly not necessary to read these notes to enjoy the book, but they may help readers to better understand some of the concepts presented in these pages.

Aspect's experiments

The experiments conducted by Alain Aspect and his collaborators (Philippe Grangier, Gérard Roger and Jean Dalibard) at the Université d'Orsay in France marked an important step forward in our understanding of the quantum world by confirming the existence of correlations between spatially separated particles that had interacted in the past — correlations that could not be described by classical physics. The existence of such correlations in quantum physics, and their paradoxical nature in the context of classical (relativistic) physics, had been observed by Albert Einstein, Boris Podolsky and Nathan Rosen in 1935. In 1964, the theoretical physicist John Stewart Bell brought new light on the paradoxical existence of long-distance correlations between sub-systems with a common origin: he demonstrated the existence of a quantitative measure of these correlations ('Bell's inequalities') that could not be surpassed in the classical physics of locally interacting systems, but which could be surpassed in quantum physics.

Aspect's experiments measured the correlation between the polarizations of two photons that were spatially separated, but had been emitted in an initial state of zero total angular momentum. In conformity with the predictions of quantum physics, these experiments showed a 'violation of Bell's inequalities'; that is, a greater correlation between separate photons than what would be permitted by a classical and local description of reality. In short, Aspect's experiments confirmed the non-local nature of quantum reality. See **quantum entanglement**.

black body

In thermodynamics, a black body is an idealized physical object that absorbs all incident electromagnetic radiation. A practical incarnation of a black body is the inside surface of an oven with a small opening in one wall. All light or electromagnetic radiation that lands on that opening will penetrate into the oven and will be totally absorbed over the course of the multiple scatterings that take place on the oven's walls. When the oven's inside cavity is heated to a certain temperature T (measured in Kelvin, so that a temperature of T_c degrees Celsius corresponds to T_c + 273.15 Kelvin), equilibrium is reached between the light (in the form of radiant heat) present inside the oven and the material walls (which absorb and continuously re-emit light). In October 1900, Planck discovered (without a proof) the mathematical formula describing how the energy of the light inside of an oven (heated to temperature T) is distributed according to the light's frequency f (and thus its colour). This formula, called Planck's law of black body radiation, posits that the energy of the radiant heat present in each cubic centimetre and in each unit interval of frequency (say, one Hertz) is equal to:

$$u_f = \frac{8\pi h}{c^3} f^3 \frac{1}{e^{hf/kT} - 1}$$

where h is Planck's constant, c is the speed of light (c = 2.99792458 x 10^{10} cm/s) and k the Boltzmann constant (k = 1.380649 x 10^{-16} erg/K) which gives the energy equivalent of a Kelvin. The curve fu_f representing the energy present in each octave of frequency begins at zero (when f = 0), reaches a maximum (when hf/kT = 3.9207), and then rapidly decreases towards zero as f continues to increase. This concentration of the distribution of radiant heat around the frequency f_{max} = 3.9207 kT/h explains why the Sun (surface temperature roughly equivalent to 6,000 K) appears yellow to us, whereas burning embers (T ~ 1,000 K) appear more reddish (thus corresponding to a lower frequency). In fact, the colours seen by the human eye depend on both the light spectra of the object being observed and the retina's sensitivity to particular colours. Nonetheless, let us remember that if, all other constants remaining the same, Planck's constant were to become ten times smaller than it is, burning embers would predominantly emit light that is thousands of times brighter and centred around frequencies ten times higher (that is, in the far ultraviolet spectrum).

Bohr, Niels (1885–1962)

In 1913, Niels Bohr presented a schematic quantum model of the simplest atom, the hydrogen atom. This model combined a classical description (the electron's circular orbits around a central proton) with a number of quantum postulates (the existence of a sequence of discrete states characterized by a quantized angular momentum $L = rp = nh/(2\pi)$ where $n = 1, 2, 3...$; and by abrupt transitions between quantized states (say, m and n) with the emission of light whose frequency f_{mn} is tied to the difference of energy $E_m - E_n$ through the (Planck-Einstein) equation $E_m - E_n = hf_{mn}$. This strange combination of hypotheses led to a formula for the quantized energy levels of the hydrogen atom:

$$E_n = -\frac{1}{2}m\left(\frac{e^2}{\hbar}\right)^2\frac{1}{n^2}$$

(where m is the mass of the electron, e its electrical charge, and $\hbar \equiv h / (2\pi)$), which closely agreed numerically with the frequencies of the light emission spectra of the hydrogen atom.

Alongside Born and Heisenberg, Bohr's name is also associated with the so-called Copenhagen interpretation of quantum mechanics. This interpretation (which varies somewhat depending on which scientist's views are being considered) is based on a combination of classical physics and quantum physics, and on a statistical description of the supposedly classical results of the measurement of quantum objects. It uses Bohr's principle of complementarity (1927) between the different possible classical descriptions of microscopic quantum objects (such as the wave-particle duality which is quantitatively measured by Heisenberg's **uncertainty principle**, 1927). Starting in October 1927, when Bohr met Einstein at the 5th **Solvay Conference**, the two would embark on a years-long debate concerning the epistemology of quantum physics.

Born, Max (1882–1970)

Born made several major contributions to quantum physics. He was one of the first to see the need for replacing classical mechanics (Newtonian or relativistic) with a new 'quantum mechanics'. When on 9 July 1925, the young Heisenberg — his former assistant — showed Born the article that he had just written where he introduced a new formulation for atomic mechanics using infinite tables, Born recognized that Heisenberg's infinite tables represented what mathematicians called matrix algebra. In collaboration with his assistant at the time, Pascual Jordan, then with Heisenberg himself, in 1925 Born developed a complete version of quantum mechanics, called matrix mechanics. In 1926, after Schrödinger elaborated a second version of quantum mechanics called wave mechanics, Born studied the collision process of an electron with an atom and hypothesized that the square (of the modulus) of the electron's wave function ψ (x, y, z) must be interpreted as the probability of finding the electron at a given point (x, y, z) in space. Born summarized this statistical interpretation of quantum theory as follows: 'The motion of particles conforms to the law of probability, but the probability itself is propagated in accordance with the law of causality.' This 'law of causality' was an allusion to **Schrödinger's equation**.

Broglie, Louis de (1892–1987)

The fundamental idea behind Louis de Broglie's thesis (1924) was 'to extend to all particles the coexistence of waves and particles discovered by Einstein in 1905 in the case of light and photons'. He associated with each matter particle of mass m, of energy $E = mc^2/\sqrt{1-v^2/c^2}$ and of linear momentum $p = mv/\sqrt{1-v^2/c^2}$, a wave frequency $f = E / h$ and a wavelength $\lambda = h / p$. See **interference**.

decoherence

The concept of decoherence was introduced in 1970 by Dieter Zeh and was further developed by a number of scientists (Wojciech Zurek, Erich Joos and Dieter Zeh, Murray Gell-Mann and James Hartle) starting in the 1980s. Zeh's basic idea was an attempt to clarify Everett's vision of quantum theory by emphasizing the role played by the *coupling* of the physical system being studied (say, a radioactive atom and a cat) with its environment (say, the Geiger counter, the mechanism that unleashes the poison, the air in the box, as well as the internal microscopic configuration of the cat's body). The **wave function** of the total system is initially written as (time zero): $\psi^0_{total} = \psi_{atom}$ (not decayed) ψ_{cat} (alive) ψ_{env} (initial). After a certain amount of time has passed (corresponding to the half-life of the radioactive atom), the wave function of the total system (obtained by solving **Schrödinger's equation**) looks as follows:

$$\psi^1_{total} = \psi_{atom} \text{ (not decayed) } \psi_{cat} \text{ (alive) } \psi_{env} \text{ (atom not decayed)}$$
$$+ \psi_{atom} \text{ (decayed) } \psi_{cat} \text{ (dead) } \psi_{env} \text{ (atom decayed).}$$

This is a linear (coherent) superposition of two states of a system, but the point being made by Zeh and his successors was that the very large number of variables (not explicitly observed) contained in the wave function ψ_{env} of the environment, combined with the great difference between the two environments (where an atom decayed or not), means that the two parts written above of wave function ψ^1_{total} behave, for all practical purposes, as if they each described a different world (the first one with a non-decayed atom and a living cat, and the second with a decayed atom and a dead cat). Mathematically, the two pieces of the wave function are orthogonal to each other in a strong sense (that is, even when considering the dynamical effects of variables related to the atom and the cat, to the exclusion of variables hidden in the environment). As mentioned in the note on **Schrödinger's cat**, recent experiments have confirmed this rapid orthogonalization of two macroscopically distinct pieces of a wave function, which is called *decoherence*.

E = hf, a lesser known equation

Everyone knows (or thinks they know) Einstein's famous equation $E = mc^2$, which relates energy and mass, with *c* representing the speed of light. On the contrary, the equation $E = hf$, which connects energy and frequency, is relatively unknown, despite the fact that it is even more fundamental than $E = mc^2$. In 1900, Planck was the first to write an equation of the form $\Delta E = hf$, even if for him ΔE probably only played the role of a formal trick to segment the axis of energies into finite intervals, thus allowing him to calculate the entropy of a system of linear oscillators (of frequency *f*) that represented, in simplified form, the light-emitting atoms in the walls of an oven (see **black body**). The first physicist to give a full physical significance to $E = hf$ was Einstein, in 1905–06. More specifically, in 1905 Einstein hypothesized that a light wave (with frequency *f*) was composed of grains of light (that he called quanta), each one of these grains being localized and having the energy $E = hf$. Then, in 1906, he was the first physicist to posit that the energy of *matter* was 'quantized', that is, that it could only take certain, discontinuous values. He demonstrated that Planck's **law of black body** radiation as proposed in 1900 could only be deduced from the general laws of statistical physics if the energy of each material oscillator (present inside the oven) only took a discontinuous series of values: 0, *hf, 2hf, 3hf*....

Moreover, the equation $E = hf$ was applied by **de Broglie** in the case of the total mass-energy of a (free) material particle; it also provides the backbone for **Schrödinger's equation**.

Einstein, Albert (1879–1955)

Einstein is particularly well known for his two theories of relativity, but it is important to recall that, as Max Born explained, 'Even if Einstein had never written a single word on relativity, he would still be one of the greatest theoretical physicists of all time.' In particular, Born was referring to Einstein's numerous pioneering contributions to quantum physics, as seen below.

1905: Einstein introduces the revolutionary idea that light is composed of packets of energy ($E = hf$).

1906: 1) March: Einstein proves that the energy of a material oscillator must take quantized values (0, hf, $2hf$, $3hf$... etc.)

2) November: He demonstrates that the quantization of the energy of the material oscillators explains the 'abnormal' behaviour of the specific heat of certain bodies (diamonds in particular).

1909: He shows that the fluctuations in the energy density of black-body radiation confirm that light simultaneously possesses wave-like and particle-like properties.

1916: 1) Einstein proves that each quantum of light emitted or absorbed by an atom during a 'quantum transition' between two possible energy levels in the atom (e.g., E_m and E_n) not only carries the energy $E_m - E_n = hf_{mn}$, but also the linear momentum $p_{mn} = hf_{mn} / c = h/\lambda_{mn}$. 2) He discovers a new quantum process: the illumination of an atom by incident radiation of frequency f stimulates the atom to effect a transition from a higher energy level, E, towards a lower energy level, $E - hf$, by emitting a quantum of light of energy hf and momentum hf/c in the same direction as the incident radiation (this process of stimulated emission is the basic concept behind how lasers work).
 3) He introduced chance into quantum physics and characterized it quantitatively by introducing several infinite tables of coefficients (A_{nm} and B_{nm}) which would provide the basis of the discovery of quantum mechanics by Heisenberg, Born and Jordan.

1924: Einstein proposes (independently from de Broglie) the attribution of wave-like properties to particles of matter; he introduces the quantum statistics of a gas of matter particles; and he discovers a new physical phenomenon, of purely quantum origin, generally called 'Bose-Einstein condensation'.

After 1927, Einstein would no longer closely follow the advances made in quantum theory. He remained unsatisfied by the statistical interpretation of the quantum formalism given by Born, Bohr and Heisenberg. He hoped to be able to 'deduce the quantum probabilities' from an underlying structure of reality, a reality which would be logically clear and would exist independently of its being observed.

In 1935, with Boris Podolsky and Nathan Rosen, Einstein pointed out the fact that there seems to be something 'spooky' about quantum theory: two particles (or two systems) that have interacted in the past preserve an intimate 'relationship' that makes every observation on a particle instantly seem to affect the other, even if they are now separated by vast distances. The discovery of such an **entanglement**, at a distance, between two systems led to many other breakthroughs and potentially heralds a new quantum revolution.

Finally, Einstein's last contribution to quantum theory was contained in the comments he made in April 1954, during the last seminar he taught at Princeton University. See **Einstein's mouse**.

Einstein's mouse

On 14 April 1954, Einstein gave the last lecture of his life to a room of sixty or so students (and a handful of professors) at Princeton University. The central theme of his presentation concerned quantum theory. Einstein explained why he thought that this theory was not the final word on the matter. He recalled in passing that while he had introduced probability in quantum physics, he was unsatisfied with the physical meaning of the wave function. He gave the example of the quantum description of a small, 1mm-diameter ball moving around in a box. If one waits long enough, the wave function gives a nebulous description of the ball's position in the box, whereas everyday experience tells us that we always see the ball in a specific spot. He thus added that 'it is difficult to believe that this description is complete. It seems to make the world quite nebulous unless somebody, like a mouse, is looking at it. Is it believable that the gaze of a mouse can drastically alter the universe?'

Einstein's colourful language struck the young Hugh Everett, who was in attendance with his friend Charles Misner. He quotes Einstein in the long version of his thesis, commenting that from the standpoint of his theory, 'it is not so much the system which is affected by an observation as the observer, who becomes correlated to the system', and that 'the mouse does not affect the universe — only the mouse is affected.'

Everett, Hugh (1930–1982)

When he was a young student at Princeton, Hugh Everett attended Einstein's last seminar (see **Einstein's mouse**) and was intrigued by the latter's remarks on the apparently incomplete nature of quantum theory, which offers a 'nebulous' description of the universe and which seems to require the presence of living beings, even just a single mouse, to set off what supporters of the Copenhagen interpretation call 'wave function collapse', i.e., the shift from a nebulous quantum world (described by the wave function ψ) to the sharp world that we see all around us (where cats are either dead or alive, but do not exist in a superposition of both states). A few months later, in the fall of 1954, Niels Bohr visited the Institute for Advanced Study in Princeton and gave a lecture on quantum mechanics where he argued that his idea of 'complementarity' solved the problem of wave function collapse. This argument seemed absurd to Everett. Not long after, during a particularly boozy evening, a heated discussion of the interpretation of quantum theory took place at the Graduate College between Hugh Everett, Charles Misner and Aage Petersen, Bohr's young assistant. In the heat of the discussion, Everett had a first glimpse of his new understanding of the quantum universe, which he presented to his friends: the universal wave function.

This idea consists in stating that there is no wave function collapse, no mysterious transition from a nebulous quantum world to a sharp classical world, but that quantum reality is simply defined by the wave function ψ_{total} of the total system, which includes not only the atomic sub-system being considered, but also the measuring instruments, the observers looking at the results on the said instruments, and even the memory of these observers with regard to previous results. Following Bryce DeWitt, who was with Dieter Zeh the first physicist to take Everett's idea seriously, this notion has often been described as the 'many-worlds theory'. But this was not how Everett characterized his idea. At the most, he spoke of the 'multiplication of observers', having in mind the idea that the wave function ψ_{total} after measurements on an atomic sub-system is a linear superposition of several sub-wave-functions $\psi_{partial}$ each evolving independently of each other and each one describing the vision and memory of what can be called a 'definite observer'. See **decoherence**.

After hitting upon this idea and beginning to develop it mathematically, Everett went to see John Wheeler and asked him to be his doctoral thesis adviser. Wheeler accepted, which was both a blessing and a curse for Everett. A blessing, because Wheeler was very open to new ideas and encouraged his students to think for themselves. A curse, because Wheeler had unconditional admiration for Bohr and his complementarity principle. Ultimately, following Wheeler's advice, Everett did not publish the long text outlining his idea, but instead wrote a much shorter (and much less clear) text that he defended as his dissertation in 1957 and which was published that same year.

Despite — or no doubt because of — its originality, Everett's interpretation did not attract much interest at the time, except for a smattering of criticism (and underhanded attacks) by physicists associated with Bohr. The importance of the new way forward signposted by Everett was only recognized during the 1970s, mainly through the work of Dieter Zeh and Bryce DeWitt. Today, according to a recent survey conducted via email, the majority of theoretical physicists working in **quantum cosmology** adopt Everett's interpretation, albeit sometimes with some personal variations. Nevertheless, if we look at the much larger community of scientists (theoretical and experimental) working in or with quantum physics, only a small minority is ready to espouse the Everettian vision of the universal wave function.

Heisenberg, Werner (1901–1976)

On 7 June 1925, suffering from an acute case of hay fever, the young (23-year-old!) Werner Heisenberg left Göttingen (where he was Max Born's assistant) to take refuge on the island of Heligoland, in the North Sea, where there was little pollen in the air. It was on this island that he discovered the first complete formulation of quantum mechanics. Inspired by his previous work with Hendrik Kramers (work that had drawn on the infinite tables A_{mn} and B_{mn} introduced by **Einstein**), Heisenberg discovered a new quantum dynamics according to which the equations describing the classical dynamics of the position q and momentum p of an electron were preserved as is, but where the classical variables q and p were replaced with two infinite tables, $[q]$ (composed of q_{mn}) and $[p]$ (composed of p_{mn}). He used physical reasoning to construct the product of two such infinite table and found that the product $[q][p]$ could not be equal to the product $[p][q]$ and that their difference was equal to an infinite table all of whose elements were of zero value, except those for which $m = n$ and which were equal to $ih/(2\pi)$, where i was the imaginary square root of -1. This highly abstract 'matrix mechanics' was further developed by Max Born, Pascual Jordan and Heisenberg himself upon the latter's return to Göttingen at the beginning of July.

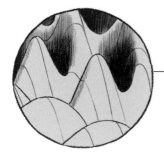

interference

When two travelling waves of the same frequency and same wavelength overlap with each other (for example, behind a double slit illuminated by a single incident wave), the superposition of the two waves leads to a total oscillation whose amplitude differs throughout space. In some places, the two waves add to each other, while in others they may partially cancel each other out (they may completely cancel each other out if both have the same amplitude). The spatial variation in the wave's amplitude obtained by the superposition of two waves is called interference. In classical physics, the observation of this kind of phenomenon (for example, in the case of a double slit illuminated by an incident light wave) was viewed as proof of the fact that light was a wave. In quantum physics, however, if electrons are sent *one by one* towards a sufficiently narrow slit, interference occurs between the partial wave functions created (when exiting the slits) by the wave ψ (**de Broglie's** wave) describing the electron's dynamics. Consequently, the amplitude of existence (**Everett**) and thus the probability of detecting the electron (**Born**, as interpreted by Everett) will vary greatly in the space behind the double slit. Phenomena of this sort were observed (with electrons) as early as 1927 by Clinton Davisson and Lester Germer, which led to Louis de Broglie winning the Nobel Prize in physics in 1929.

multiverse

The term 'multiverse' is used in a variety of different contexts. It is sometimes used to refer to the 'quantum multiverse' envisioned by **Hugh Everett**, i.e., the multiplicity of the quantum world as described by the universal **wave function**. But the word is also used to designate the essentially classical multiplicity of slightly (or greatly) different local universes that are generated by certain kinds of primordial cosmological inflation (the 'eternal inflation' model discovered by Alexander Vilenkin in 1983).

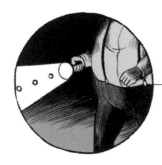

photon

A photon is the term now used to refer to what Einstein introduced in 1905 as a 'light quantum'. It designates the quantum particle associated with the electromagnetic field. Its energy and momentum are given respectively by $E = hf$ and $p = hf/c = h/\lambda$, where f is the frequency of the associated light wave and $\lambda = c/f$ is its wavelength. The 'invariant mass' (in the relativistic sense) of a photon [that is to say the square root of $(E/c^2)^2 - (p/c)^2$] is zero.

Planck, Max (1858–1947)

In 1899–1900, Max Planck founded quantum theory when he discovered the constant h that is now named after him (see **Planck's constant**). Most likely, at the time he did not understand the exact role that h played in the physics of the linear oscillators that he used to model the matter of the walls of a **black body**.

In fact, at the beginning Planck essentially used $\Delta E = hf$ as a mathematical device allowing him to count, as Ludwig Boltzmann had already done, the number of ways of distributing the total energy among the different oscillators (of the same frequency). However, he clearly understood from the beginning that the discovery of this new universal constant h was (as he told his son at the time) 'a discovery of the highest order, only comparable perhaps to Newton's discoveries'. Between 1900 and 1913, Planck tried to expand the significance of the constant h. In particular, he emphasized the fact that h has dimensions of physical 'action' (obtained by multiplying an energy by an interval of time, or a momentum by an interval of distance, then by adding up the intervals of time or space). The role of h as a quantum of action played a very important part in a number of subsequent developments in quantum physics, in particular those spurred on by Louis de Broglie, Paul Adrien Maurice Dirac and Richard Feynman.

Planck's constant: h

Planck's constant, h, was introduced to physics by **Max Planck** in 1899–1900 in the context of his theoretical research on the distribution of frequencies of radiant heat contained in an oven (see **black body**). Its value is $h = 6.626070 \times 10^{-27}$ erg s, where an erg (= 1cm^2 g/s^2) is the unit of energy in the CGS (centimetre, gram, second) system of units. It is often useful to replace it with the 'reduced Planck constant', $\hbar \equiv h / (2\pi) = 1.0545718 \times 10^{-27}$ erg s.
Planck's constant embodies quantum physics. It is essential to all of the equations and physical laws related to quantum theory: the Planck-Einstein relation $E = hf$ establishing the proportional relationship between energy and frequency; de Broglie's equation $p = h/\lambda$ associating momentum and wavelength; Heisenberg's commutation relation $\hat{q}\hat{p} - \hat{p}\hat{q} = i\hbar$ between position and momentum operators; **Schrödinger's equation**, and so on. Classical physics is formally obtained by taking the limit where h approaches zero. But it is important to note that the structure of the world around us is largely dependent on the small but non-zero value of h. See **the quantum universe and everyday life**.

quantum cosmology

Cosmology is the scientific study of the whole universe. When adopting the Copenhagen interpretation of quantum mechanics, which hypothesizes that the macroscopic world of the objects around us is still described by classical physics, it seems absurd to talk about quantum cosmology. Nonetheless, motivated by the need to reconcile Einstein's theory of general relativity (which is essential to cosmology) and quantum theory, a number of researchers (Hugh Everett, Bryce DeWitt, Stephen Hawking, Alexei Starobinsky, Viatcheslav Mukhanov) have explored the consequences of quantum theory on cosmology. One of the most remarkable findings (obtained by V. Mukhanov and G. Chibisov in 1981) was discovering the *quantum* origin of the density fluctuations in the primordial universe that gave rise to the structure of the universe as we know it. For example, our galaxy, with its hundreds of billions of stars, is the product of a primordial quantum fluctuation. Many researchers working in this field are following in the footsteps of Hugh Everett, who presented one of the only clear avenues for thinking about quantum reality.

quantum entanglement

Quantum entanglement is the term used to describe the correlation that exists (in certain situations) between two (or more) quantum objects, regardless of the distance separating them. Technically, this correlation stems from the fact that quantum theory defines the state of a composite system $A + B$ by means of a single wave function ψ (q_A, q_B) that depends on the variables that describe both the configuration of object A and that of object B. Except in the very specific case where the total wave function ψ (q_A, q_B) can be written as the product of a function depending only on the variables q_A describing A and of another function depending only on the variables q_B describing B, the predictions made based on the total wave function ψ (q_A, q_B, t) will imply correlations between the observations that one can make (at instant t) of A and the observations that one can make (at the same instant t) of B. These correlations continue to exist even when the sub-systems A and B are separated by a great distance. The existence of quantum entanglement was first signalled by Albert Einstein, Boris Podolsky and Nathan Rosen in 1935. The first entirely convincing experimental proof of the existence of quantum entanglement (in a situation where the correlations due to quantum entanglement were stronger than anything that could be explained by classical physics) was obtained in **Aspect's experiments**.

the quantum universe and everyday life

Alongside the theory of special relativity and the theory of general relativity, quantum physics is one of the cornerstones of our current understanding of the laws of nature. Recent developments in cosmology and in the lab have confirmed that the quantum description of reality operates not only at the atomic level, but also at the level of macroscopic objects, including at the level of the universe as a whole. Quantum physics is all around us: fusion reactions within the sun, the reddish colour of burning firewood, the solidity of the chair that we're sitting on, the characteristically yellow colour (also lighting up our city streets) that occurs when we throw a bit of salt on a fire, and so on.

It also provides the foundation for many technological applications that have shaped our modern society: lasers (and hence CDs, DVDs, the internet's fibre-optic amplifiers), microprocessors (and thus computers, smartphones, the internet), atomic clocks (and thus GPS...), magnetic resonance imaging devices, etc. Quantum physics is also the backbone of chemistry. And not only has chemistry shaped our daily lives, it is at the root of the existence of all life (in particular through the stability and coding potential of DNA molecules). It has also been suggested that the intimate nature of the quantum universe, made of the coherent superposition of different classical realities, plays a useful role in several biological structures: the light receptors of photosynthetic organisms, the magnetic receptors of certain migratory birds...

Schrödinger, Erwin (1887–1961)

At the end of 1925, guided by the work of Einstein and de Broglie on the quantum description of perfect gases (which exhibited wave-particle duality) and unsatisfied by the highly abstract nature of the matrix mechanics developed by Heisenberg, Born and Jordan, Erwin Schrödinger decided to pursue the path laid out by de Broglie by writing a partial differential equation describing the propagation (over time) in the configuration space of the system in question of a wave that he denoted $\psi(q, t)$. See **Schrödinger's equation**. In a series of articles published in quick succession at the beginning of 1926, he expounded upon the many implications of his equation. He also demonstrated the equivalence between his 'quantum wave mechanics' and the 'quantum matrix mechanics' developed by Heisenberg, Born and Jordan. But he also observed the strange fact that the wave function $\psi(q, t)$ does not describe a set of several quantum particles by using several different waves, propagating in the usual three-dimensional space, but that it in fact uses a single wave propagating through the abstract space descring simultaneous configurations of the set of all the particles in question.

The strange representation of reality offered by the wave function would trouble Schrödinger for the rest of his life. Like Einstein, he believed that the 'tranquilizing philosophy of Heisenberg and Bohr', i.e., the Copenhagen interpretation of quantum theory, left a number of legitimate questions about quantum reality unanswered. See **Schrödinger's cat**.

Schrödinger's cat

In 1935, to illustrate a situation predicted by quantum theory and which seemed absurd from a classical perspective, Schrödinger posited the case of a cat living inside of a box that also contained a diabolical mechanism that might or might not kill the cat within an hour's time, depending on whether a single radioactive atom decayed or not. Quantum theory describes the cat at the end of an hour using a wave function ψ, which corresponds to the equal superposition of the wave ψ_{alive} for a cat that is still alive and the wave ψ_{dead} for a cat that is dead: $\psi = \psi_{alive} + \psi_{dead}$. With this example given in passing, Schrödinger wanted to illustrate that it was impossible to limit 'quantum uncertainty' to the microscopic world. In the same article, Schrödinger also introduced the word 'Verschrankung', or **entanglement**, to describe the kind of situation previously studied by Einstein, Podolsky and Rosen (EPR).

For a long time, 'Schrödinger's cat' remained a mere thought experiment, used to illustrate the paradoxical nature of the quantum description of reality. Recently, however (in 1996), a French team led by Serge Haroche, Jean-Michel Raimond and Michel Brune conducted a series of experiments that acted out situations like 'Schrödinger's cat'. Their 'cat', which was an electromagnetic field trapped inside of a cavity, was prepared in a state of coherent superposition between two macroscopically distinct states. They observed that, because of its coupling with the environment, it very rapidly became impossible to distinguish this coherent superposition from an incoherent superposition where each different macroscopic state 'leads its own life' independently of all others. See **decoherence**.

Schrödinger's equation

The Schrödinger equation is essentially an application of the Planck-Einstein relation $E = hf$, or its reverse $hf = E$, to the evolution of the wave function $\psi(q, t)$ in time t and in the space defined by the configuration variables q. In fact, it can be written as follows:

$$h\hat{f}\psi = \hat{E}\psi$$

where $\hat{f}$ and $\hat{E}$ denote certain operations conducted on the wave function ψ. The operation $\hat{f}$ acting on ψ designates the 'operator' $\frac{i}{2\pi}\frac{\partial}{\partial t}$ where i is the imaginary square root of -1 and $\partial/\partial t$ is a partial derivative with respect to time. It is thus defined, so that $\hat{f}$ acting on a monochromatic wave of frequency f [that is, containing the exponential factor $\exp(-2\pi ift)$] simply gives the product of f and of the wave in question. The operation $\hat{E}$ is obtained by expressing first the energy E of the system in question as a function of positions q and momenta p, $E = H(p, q)$, then by replacing each momentum p with an operation of partial differential with respect to the corresponding position q: $\hat{p} = -\frac{ih}{2\pi}\frac{\partial}{\partial q}$. The operation $\hat{p}$ acting on a travelling plane wave of wavelength λ [that is, containing an exponential factor $\exp(2\pi iq)/\lambda$] simply gives the product of h/λ and of the wave in question.

Since, as de Broglie proved, h/λ is equal to the momentum p of the particle associated with the wave, we can see that the operation $\hat{p}$ ψ simply gives the product $p\psi$. As a result, the operation $\hat{E} = H(q, \hat{p})$ acting on a plane wave gives us back the product $E\psi$. Essentially one can say that Schrödinger found his equation by decomposing a wave of very general shape (satisfying an equation of linear propagation) into monochromatic plane waves (Fourier series decomposition) and by applying the Einstein-de Broglie relations $E = hf$, $p = h/\lambda$ to each element of this decomposition. After he discovered this equation in late December 1925, Schrödinger was able to solve it for the hydrogen atom [that is, for a single electron, with $q = (x, y, z)$, whose energy is the sum of its kinetic energy $p^2/2m$ and the energy of attraction of the central proton $- e^2/|q|$]. He thus found that the only possible values for the energy of the hydrogen atom were those heuristically deduced from Bohr's model.

Solvay Conferences

The Solvay Conferences periodically bring together experts in a specific field to discuss outstanding scientific questions. The first Solvay Conference took place from 30 October to 3 November 1911, at the Hotel Metropole in Brussels; its theme was 'The Theory of Radiation and the Quanta'. It was presided over by Hendrik Lorentz and was the first international conference that Einstein attended. With regard to quantum physics, one of the most significant Solvay Conferences occurred in 1927, when all the different generations of pioneering physicists in quantum theory were present: Planck, Heisenberg, Einstein, Bohr, Born, de Broglie, Schrödinger, Dirac and Pauli. The discussions that took place between Bohr and Einstein during this conference have been the subject of much subsequent commentary.

The Solvay Conferences continue to this day, still taking place at the Hotel Metropole, under the aegis of the International Solvay Institutes for Physics and Chemistry, headed by Marc Henneaux, with a view to furthering cutting-edge physics, astrophysics and chemistry. In 2011, the 25th Solvay Conference focused on the theme of 'The Theory of the Quantum World'. This is the session of the Conference that our hero Bob attends.

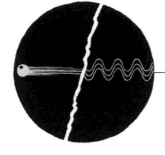

uncertainty principle

In 1927, while he was in Copenhagen with Bohr and recalling a suggestion of Einstein's, Heisenberg understood that the fact that the difference $[q][p] - [p][q]$ between the product of his tables (which, according to Schrödinger's formalism becomes the difference $q\hat{p} - \hat{p}q$ where $\hat{p}$ is the operation defined in **Schrödinger's equation**) was non-zero (and equal to $ih/2\pi$)), implied a *principle of uncertainty* regarding the simultaneous measurement of the position and momentum of a particle. This principle concerns the uncertainty Δq with which one can measure the position q and the uncertainty Δp, with which one can measure the momentum p and states that their product $\Delta q \Delta p$ is necessarily greater than $h/(4\pi)$. Bohr then suggested that this mathematically accurate result be interpreted in terms of 'complementarity' between the particle (q) and wave-like ($p = h/\lambda$) aspects of any quantum object.

wave function, ψ

Classical physics describes a physical system by representing it mathematically with a certain number of configuration variables, i.e., q. [Here, the letter q stands for a long list of individual variables. For example, for a system with two particles, q represents a list of six variables: the three spatial coordinates (x_1, y_1, z_1) of the first particle and the three spatial coordinates of the second particle (x_2, y_2, z_2).] In classical physics, the evolution of the system over time, $q(t)$, is fully determined by the initial values of the configuration variables q_0 at a certain initial moment (e.g., $t = 0$), and by their initial 'velocities' $\dot{q}_0$ (that is, their time derivative at the initial moment). It is often helpful to use the so-called Hamiltonian formulation of classical physics where the velocities $\dot{q}$ are replaced by certain 'conjugate momenta' p. [For a non-relativistic particle, $p = m\dot{q} = mv$ describes the particle's momentum].

According to Schrödinger's interpretation, quantum physics describes the same physical system using a complex function of

q and of time t, $\psi(q, t)$, called the system's 'wave function'. $\psi(q, t)$ measures the 'amplitude of existence' of a given system in the configuration classically described by q. In other words, while in classical physics a given system exists, at each instant, in a single, well-defined configuration $q(t)$, in quantum physics this system 'exists' at each instant simultaneously in a superposition of an infinite number of different configurations, each configuration q 'existing' with a certain (complex) amplitude $\psi(q, t)$. This kind of 'delocalized' existence of different classical configurations is represented, in this text, by superposing several images of different colours and intensities (by using the modulus ρ of the complex number $\psi = \rho e^{i\theta}$ to measure the intensity (brightness) and its phase θ to encode a shade on the colour wheel). The evolution over time of wave function $\psi(q, t)$ is given by **Schrödinger's equation**.

Mathieu Burniat would especially like to thank Jérôme Loreau, who opened the door to physics for him. He would also like to thank Emma, Nicholas, Anne, Antonio, Tiago, Baudouin and Samuel for their invaluable support.

Thibault Damour is grateful to Bryce DeWitt, Slava Mukhanov and Dieter Zeh for illuminating discussions on the quantum multiplicity of the world. He would also like to thank Charlie Misner and Hale Trotter for their invaluable first-hand information regarding Hugh Everett, and A. Douglas Stone for his kind and expert help in reviewing the English translation.

The authors thank Pauline Mermet and the whole Dargaud team for the enthusiasm with which they supported this quantum adventure.

PENGUIN BOOKS

UK | USA | Canada | Ireland | Australia
India | New Zealand | South Africa

Penguin Books is part of the Penguin Random House group of companies whose addresses can be found at global.penguinrandomhouse.com.

Penguin
Random House
UK

First published in French by Dargaud 2016
This translation first published by Particular Books 2017
Published in paperback in Penguin Books 2020
003

Printed and bound in Italy by L.E.G.O. S.p.A.

A CIP catalogue record for this book is available from the British Library

ISBN: 978-0-141-98517-6

www.greenpenguin.co.uk

MIX
Paper from
responsible sources
FSC® C018179

Penguin Random House is committed to a sustainable future for our business, our readers and our planet. This book is made from Forest Stewardship Council® certified paper.